INTERNATIONAL CENTRE FOR MECHANICAL SCIENCES

COURSES AND LECTURES - No. 45

DONALD BAIN

BRITISH HYDROMECHANICS RESEARCH ASSOCIATION

HEAVY CURRENT FLUIDICS

COURSE HELD AT THE DEPARTMENT
OF FLUIDDYNAMICS
OCTOBER 1970

UDINE 1970

SPRINGER-VERLAG WIEN GMBH

ISBN 978-3-211-81148-1   ISBN 978-3-7091-3000-1 (eBook)

DOI 10.1007/978-3-7091-3000-1

PREFACE

The disclosure by the Harry Diamond Laboratories, in 1960, of the principles of the first fluid amplifiers gave recognition to a new technology known variously as fluidics, fluerics or fluid amplification.

Interest in fluidics has grown steadily throughout the last decade and international conferences and symposia are devoted solely to the technology.

So far the majority of effort has been concentrated on small logic devices and systems have been built which contain several hundred elements.

Research and development at the British Hydromechanics Research Association, Cranfield, England, was concentrated however on studies of large elements which controlled the working fluid itself, the nozzles of such elements may be measured in metres rather than millimetres. The primary purpose of this text is to point out and explain the relevant physical phenomena employed in the different types of fluidic device developed at the British Hydromechanics Research Association. Typical performance figures are given together with examples of the industrial application and use of these elements either individually or in systems. Throughout, emphasis has been made on the development of devices for industrial use.

D.C. Bain

1. Introduction

Fluidics is a new technology which utilises the flow of fluids to accomplish sensing computation or control. Although preliminary investigations were made on some fluidic elements at the beginning of the 20th century the recognition of fluidics as a new technology is usually attributed to the announcement in 1960 of the work being carried out at the Harry Diamond Laboratories in the U.S.A.

Most research and development, so far, has been on control systems with air as the fluid using digital elements of relatively small size [1]. But, by using large elements the control of the working fluids themselves can be effected. This field has come to be known as "heavy current fluidics".

We will confine ourselves to a study of pure fluidic elements which have no moving parts and use fluid jets to produce the required function.

2. THE ADVANTAGES AND DISADVANTAGES OF HEAVY CURRENT FLUIDICS

2.1 Advantages (Table 1)

Table 1

POTENTIAL ADVANTAGES AND DISADVANTAGES OF FLUIDICS

	Versus Mechanical or Electromechanical Devices	Versus Electronic Devices
Advantages of Fluidics	Vibration, shock resistant	Vibration, shock resistant
	No wearing parts	Radiation resistant
	Faster operating speed	Will not burn out or short out
	Lower cost	
	Safety in explosive environments	Safety in explosive environments
	Reduced volume, weight	
	Higher temperature capability	Higher temperature capability
	Elimination of interfaces	

Disadvantages of Fluidics	Requires fluid supply	Higher cost in some applications
	Higher power consumption	Requires fluid Supply
	Some perfor-mance change with temperature	Higher power consumption
	Interface problems (in some appli-cations)	Lower operat-ing speed

The absence of moving parts which can wear is a big advantage particularly in oscillators or other applications requiring frequent switching of the flow. Fluidic elements should therefore have a longer more maintenance free life with considerably simplified sealing problems. Wear of the flow pas-sages could occur when abrasives are present in the flow considerable geometry changes are however usual-ly required before a properly designed fluidic valve will fail and operation can usually continue until the valve wall is worn through and an external leak occurs.

Fluidic valves would appear particular-ly suited for handling slurries, provided the solids

size is small enough to preclude the possibility of
blockage. In particular, since the fluid is in continuous motion within the valve, suspended solids do
not have an opportunity to settle out. As fluidic
diverters can be switched with either liquids or
gases the control problem is transferred to one of
simple liquids at much smaller flows.

For example the pilot valves need not be subject to
the hazards of slurry service as any clean fluid can
be used for the control flow. If a small control flow
causes a dilution problem short duration control
pulses can be used to switch two position wall attachment diverters, in particular a pulse of less than a
cupful of water is sufficient to switch a diverter
passing 150 gpm.

Fluidic valves would appear suited
for extreme environments in particular the absence
of valve stem packing and moving parts should be
ideal for high temperature applications. Extremes
of vibration have little or no effect on performance,
wear and life. Since they are not tolerance dependant
a wide choice of materials and manufacturing tech-
niques can be used.

Fluidic valve switching speed is much superior to
that of conventional valves with reduced risk of

surges, small pilot valves can be used and there is
no longer any requirement for large actuators which
need time to be stroked through an appreciable travel.
Less than 120 milliseconds is required for complete
diversion of 750 g. p. m. water flow.

2.2 Disadvantages

The two outlets of a fluidic valve can
never be isolated from one another and the discharge
through the active limb depends on the differential
pressure across the outlets. There is only one set of
operating conditions for which there is no flow in
the passive limb, the pressure recovery is then a
maximun. Under other conditions there will either be
induced flow into or spill out of the passive limb.
Adjustment of the outlet loading may help but does
not eliminate the problem. Conventional non return
valves can be used to isolate the two outlets but
their inclusion removes the main advantage of the
non moving part feature of the fluidic valve.

Fluidic valves are most suited therefore to those applications where isolation of the
outlets is not essential.

Conventional wall attachment amplifiers

do not operate particularly well with two phase flow
e.g. with water as the driving fluid with a free discharge to atmosphere. In the vertical position gravitational forces will oppose the attachment forces and
the valve will not function satisfactorily at low input heads due to spill over. In the horizontal, or
near horizontal condition, fluctuating output can occur due to air breaking into the attachment bubble.

Wall attachment diverters will not func
tion down to zero flow since they depend on the exist-
ence of flow for wall attachment. In closed systems
the minimum operating flow occurs near the transition
from laminar to turbulent conditions since wall attach
ment is poor with laminar flow. In order to operate
the valve successfully down to these conditions the
outlet pressure must decrease proportionately with the
input pressure. This is more or less true if the output
pressure is due to frictional losses in the piping downstream of the outlet. However, if the outlet is discharging against some static head, then the minimum
operating flow will occur when the pressure is high
enough to overcome this head plus any pipe friction
losses at that flow.
Since fluidic valves do not work well under laminar
conditions the working fluid viscosity must not be

high enough to give laminar flow. The larger the valve
and the higher the input head then the more viscous
can the working fluid be.

3. WALL ATTACHMENT AMPLIFIER

3.1 Principle of Operation

The wall attachment amplifier makes
use of the entrainment characteristics of a submerged
fluid stream. Fig. 1 shows a subsonic jet diffusing as
it issues from
a nozzle. The ef
fects of a fluid
stream flowing
between bound-
ary walls and
emerging into
a similar medium
are shown in
Fig. 2 (see page
12). The issu-
ing jet entrains
some of the sur
rounding fluid, lowering the pressure in the zone between the stream and the boundary walls, inducing a

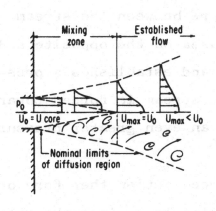

Fig. 1. Diffusion of a Jet From a Slit with
Typical Velocity Contours

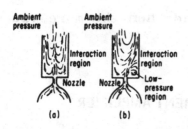

Fig. 2. (a) Initial Fluid Flow Between Parallel Walls;
         (b) Final Fluid Flow Between Parallel Walls.

flow from the higher pressure region beyond the opening to maintain equilibrium. When the stream is closer to one wall than the other, the area available for counter flow on the near side is decreased while that on the far side is increased. A decrease in area impedes the counter flow and results in an even lower local pressure between the stream and the wall. The increased area on the opposite side facilitates the counter flow and establishes a pressure approaching ambient. The stream is forced towards the closer wall resulting in an even greater pressure differential.

A self reinforced action therefore occurs shifting the stream completely against the wall. The stream constantly entrains fluid from the low pressure region, the losses being replenished by recirculation from the jet stream.

Typical flow through a wall attachment amplifier is shown in Fig. 3 (see page 13). There is

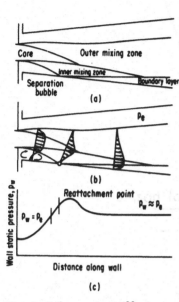

Fig. 3 Schematic Diffusion of
Attached Jet. p_e = exit.

a central core with an outer
mixing zone similar to that
for a submerged stream not
affected by walls. There is
a separation bubble near the
wall consisting of flow originally in the power stream.
A vortex motion occurs in
the separation bubble as the
stream continually entrains
fluid which is replaced by a
back flow from the attachment
point. Between the separation bubble and the core is
an inner mixing zone which
consists of flow originally in the power stream and
fluid entrained from the separation bubble.

To shift the stream from one wall to
the other control orifices are introduced on the
sides of the power stream Fig.4 (see page 14). If
control fluid is allowed to enter the separation bubble at a greater rate than the stream can entrain and
remove it the pressure will increase and the stream
will move towards the centre line. Control fluid
need only be injected until the power stream crosses

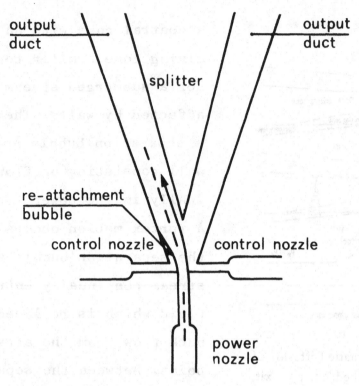

Fig. 4 Wall Attachment Diverter

the centre line when the self reinforced action occurs and the stream attaches to the other wall. Thus
by merely inhibiting entrainment, or providing the desired entrainment, the direction of the submerged
stream can be controlled.

3.2 Geometry

The typical layout of a wall attach-
ment diverter is shown in Fig. 5 (see page 15). The

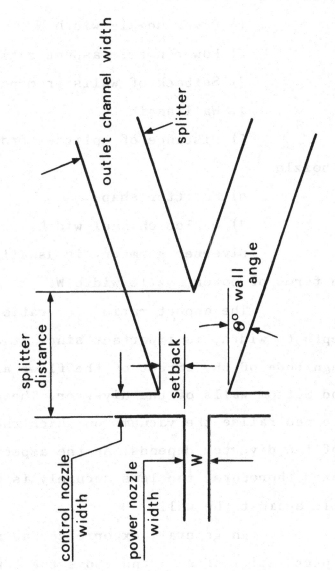

Fig. 5 Wall Attachment Diverter - Main Geometric Variables

main geometric variables are : -

        1) Power nozzle width W

        2) Power nozzle aspect ratio

        3) Setback of walls from power nozzle

        4) Wall angle

        5) Distance of splitter from outlet of
power nozzle

        6) Splitter shape

        7) Outlet channel width.

Diverter geometry is usually expressed in terms of power nozzle width W.

The aspect ratio i.e. ratio of power jet depth to width, is important since it determines the magnitude of the effect of the flows along the top and bottom walls of the diverter. These flows tend to neutralise the vacuums on which the operation of the diverter depends. As the aspect ratio is decreased therefore, the less securely is the power jet held against the wall.

In general, experience has shown that for aspect ratios of 4 : 1 and above the effect of the flows along the top and bottom walls is very small and diverter characteristics are independent of aspect ratio.

At low set backs a small switching

flow is required because of the close proximity of the power jet to the opposite wall. Low setbacks therefore give the best performance in terms of flow gain usually however at the expense of stability.

The effect of variation in wall angle on performance has not so far been thoroughly investigated and most workers use a value between 12° and 15°. In general, very small angles make it difficult to maintain bistability, on the other hand large wall angles make it difficult to maintain wall attachment.

Short splitter distances, of less than 3 nozzle widths, make the diverter very load sensitive since any pressure rise in the output limb acts immediately on the wall attachment bubble. Sufficiently large splitter distances give the diverter blocked load stability Fig. 6 (see page 18). In this case blocking the output limb of the diverter will cause the flow to remain attached to the same wall although it passes out through the other limb. On removal of the load the flow will revert to the original output limb. As an example, for a diverter with a pointed splitter, having a setback of 0.25 W, the minimum splitter distance for blocked load stability is around 6 W.

Splitter shape has an important effect

blocked output   unblocked output

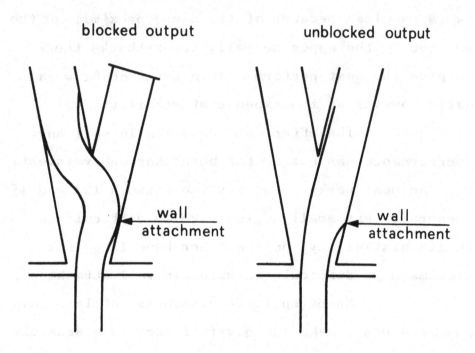

Fig. 6 Blocked Load Stability

on diverter performance. With small wall angles and
setbacks then a near pointed splitter has to be used
to maintain a reasonable output width. The use of a
cusped splitter can give greatly improved pressure
recovery and stability together with cleaner switching.

3.3 Performance

From Fig 7 (see page 19) it is apparent
that the entire flow can be forced into the wrong out
let by completely blocking the other.

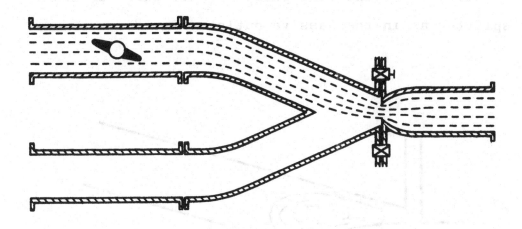

Fig. 7 Outlet Blockage

Therefore if the piping downstream of
a fluidic valve has too large a pressure drop part of
the flow will be forced into the wrong outlet i.e.
spill over will occur. If the diverter is to work properly then the outlet piping pressure drop must not
exceed some limiting value.

The flow through the valve varies as
the square root of the pressure drop from the inlet
to the splitter, the pressure drop resulting in an
increase in velocity and kinetic energy of the fluid.

With complete flow diversion i.e. with no flow in the
inactive limb, the same pressure must exist at the
splitter as in the passive outlet Fig. 8.

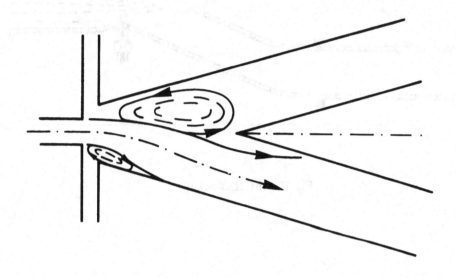

Fig. 8 Complete Flow Diversion

Thus, the fluid travelling from the power jet exit
to the splitter experiences a drop in pressure from
the inlet down to the zero flow outlet with a corresponding increase in velocity.

            In order to improve the pressure recov-
ery diffusers may be fitted in the outlet limbs of

the diverter. Experience has shown that fluidic diverters can then be made having a pressure recovery of
around 70% i.e. with an inlet pressure of 10 psig,
with the passive limb open to atmosphere, the active
limb output pressure could be as high as 6-7 psig with
complete flow diversion. The pressure recovery remains
reasonably constant over a wide range of inlet pressures.

If the resistance of the piping downstream of the active outlet is less than the maximum
allowable for complete flow diversion then fluid will
be entrained i.e. induced into the passive outlet.
Since the kinetic energy of the flow at the splitter
is not now fully recovered in pressure at the active
outlet energy is available for dissipation in entrainment.

If the resistance of the piping down-
stream of the active outlet is greater than the maximum allowable for complete flow diversion then spill-
over will occur.

A typical non dimensional flow characteristic for a wall attachment diverter is shown in
Fig. 9 (see page 22).

Switching of a wall attachment amplifier can occur as the result of the combination of

several effects. The action of flow entering through
a control port will help to satisfy the entrainment
of the power jet reducing the vacuum and strength of
wall attachment.

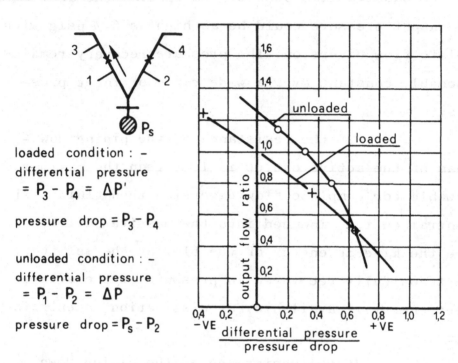

loaded condition : –
differential pressure
$= P_3 - P_4 = \Delta P'$

pressure drop $= P_3 - P_4$

unloaded condition : –
differential pressure
$= P_1 - P_2 = \Delta P$

pressure drop $= P_s - P_2$

Fig. 9 Output Loading Characteristics

Also, when the control flow possesses significant velocity, or pressure energy, it will deflect the power jet
aiding detachment. If a point is reached where the power jet strikes the splitter at an angle this causes a
low pressure region towards the vertex of the splitter
lifting the lower part of the power jet from the wall.
Where the set back is low, the jet may also be deflected

far enough to allow entrainment to cause a low pressure region on the opposite wall which can lead to complete switching. Liquid diverters are in many respects similar to pneumatic diverters when operating at the same Reynolds Number with single phase flow, there are however several important differences. The fluid velocity is generally lower in the liquid diverter, its physical size is larger and switching times are therefore much longer. Bleeds or vents are normally necessary with pneumatic diverters in order to decouple the load. Using vents, when the element is loaded the back pressure does not reach the separation bubble and any flow which cannot egress through the loaded output passes out through the bleeds. Since no flow is spilled to the passive output the outputs are therefore decoupled. Some flow is often diverted back into the interaction region forming a vortex aiding stability. With a liquid diverter it is not usually practical to vent the working liquid which would either have to be collected or wasted. With liquids, loads are mainly inertial due to the higher accoustic velocity and longer switching times and blocked load stability is usually sufficient with a liquid diverter. Care must also be taken in operating liquid diverters to avoid cavitation.

3.4 Industrial Use

The wall attachment amplifier can be
used as a flow diverter in a wide variety of industrial processes including liquid level control, liquid
transfer e.g. tank filling and draining, in sampling
systems and for self cleaning filtration systems.

Fig. 10 (see page 25) shows a typical
wall attachment diverter developed and manufactured
by B.H.R.A. for use in the chemical industry. The
diverter is both simple and cheap to manufacture using normal workshop facilities. The diverter can be
switched by either providing a positive pressure
liquid or gas flow into the appropriate control port,
or, alternatively designing the valve to give sub-
atmospheric pressure conditions at the control ports
so that merely opening the port to the atmosphere
causes switching to occur. It is the latter switching
mechanism which is used in the liquid level controller
developed by B.H.R.A.

a) Liquid Level Control

For level control of a tank with a
free liquid surface it is more convenient to operate
with only one switching signal ; therefore using asym-

Fig. 10 Wall Attachment Diverter

metrically placed side walls the bistability of the
diverter valve is converted to monostability giving
the valve a preferred output, the by-pass position of
Fig. 11 on the side with the smaller offset, which is
also the side with the control port and dip tube.

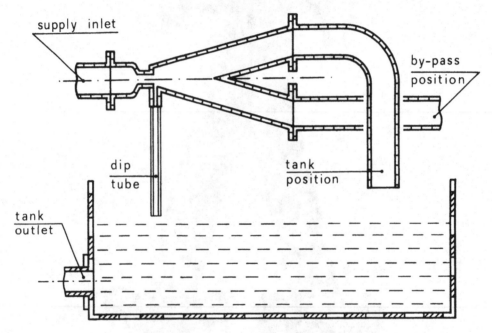

Fig. 11 Fluidic Liquid Level Control

With the dip tube open to atmosphere
i.e. the controlled liquid is low, then air entrainment through the tube is sufficient to cause the power
jet to detach from the by-pass position and switch to
tank position, thus raising the liquid level in the
free surface tank.

When the liquid just immerses the end
of the dip tube air entrainment stops and the liquid
entrainment is insufficient to maintain the power jet
in the tank filling position, thus effectively closing
the control port and causing a switch back to the by-
pass position. Hence level control is maintained by
the fluctuation of the free surface about the end
of the dip tube.

Experimental investigations show that
level control of ± 0.5 inches can be achieved quite
easily, the main source of inaccuracy being the low
rate of liquid leakage out of the dif tube as the
end is uncovered.

b) Liquid Transfer e.g. Tank Filling and Draining

Wall attachment amplifiers have been
mainly used in the process industries in systems where
the valves are completely filled with liquid when
valve characteristics are similar to those of typical
pneumatic logic devices. In some chemical plants however the systems are not filled completely the pipe-
work being designed with a slight fall, of say 1 in
20, so that the pipes drain out under gravity. One
example of this is in the transfer of the contents of
one tank by gravity flow into the receiver tanks, di-

verter valves being used to feed the flow into the
required tank, Fig. 12.

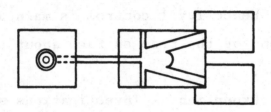

Fig. 12 Tank Filling and Liquid Transfer

Under these conditions the liquid drains out of the
unused or passive outlet of each diverter valve. The
presence of air in the interaction region also modi-
fies the valve characteristics considerably and with
the low inlet heads of this type of system gravitational effects become important. An investigation has
been carried out at B.H.R.A. into the use of fluidic
valves in low head (inlet pressures less than $41bf/in^2$)

self draining systems. At first sight the obvious
position for a self draining valve is mounted vertically but, at low inlet heads the attachment forces
on the jet are only of the same order as the gravitational forces. The component of gravitational force,
opposing attachment, reduces jet curvature resulting
in interruption of the jet by the splitter and spill
over flow. (never better than 5 % of the supply flow),
Fig. 13.

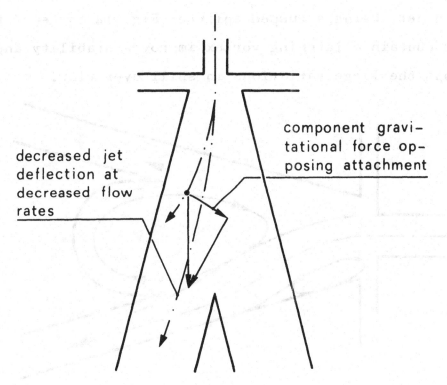

Fig. 13 Vertically Mounted Diverter

In order to overcome this disadvantage and satisfy the
self draining requirement a near horizontal position
is used.

           An important design consideration is
splitter shape. With a sharp pointed splitter and unloaded outputs a fluctuating spill over flow, as high
as 10 % of the supply flow, is obtained as the result
of jet in stability caused by the successive collapse
and restoration of the vortex on the offside of the
main jet. Using a cusped splitter Fig. 14 to generate
and contain a latching vortex improves stability and
stops the large variations in spill over flow.

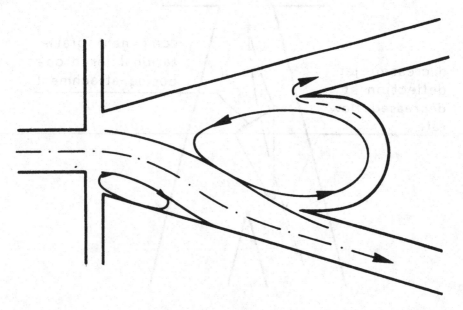

Fig. 14 Cusped Splitter

In normal fluidic elements the supply nozzle aspect ratio is greater than 2 so that shear stress effects on the top and bottom are small. In this low head application however, the gravitational effect due to channel depth exerts a considerable influence, particularly on the air water interface of the return flow from cusp. The pressure difference across the bottom of this interface causes the wall of water to spread laterally and spill over into the passive limb. Fig. 15. Minimisation of spill over by reduction in channel depth gives an optimum operating aspect ratio of unity.

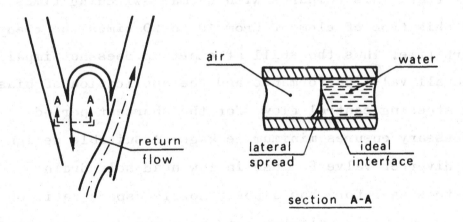

Fig. 15 Lateral Spread of Air Water Interface

At lower aspect ratios there is a marked decrease in
wall attachment giving high spillover (around 9 % of
the supply flow) throughout the pressure range.

Two aspects of transient behaviour are
important in diverter applications. First the spill
over which occurs when the valve is switched with the
power jet full established, this is minimised by correct timing of the control pulse. Second, the spill
over which occurs as the flow is being established.
Experiments show that this leakage is equivalent to
the full flow for a time of less than twice the trans
port time of the fluid between the nozzle and the
splitter. This compares with normal switching times
of this type of element from 10 to 20 times the trans
port time. Thus the spill at start up does not impair
overall valve performance and the application of bias,
or steering control flow, for the shortest period
necessary ensures minimum leakage. A suitable design
of diverter valve for use in low head self draining
systems therefore has a power nozzle aspect ratio of
unity, a cusped splitter, and is mounted in the near
horizontal position. A typical configuration and
performance of such a B.H.R.A. valve is shown in Fig.
16 (see page 33).

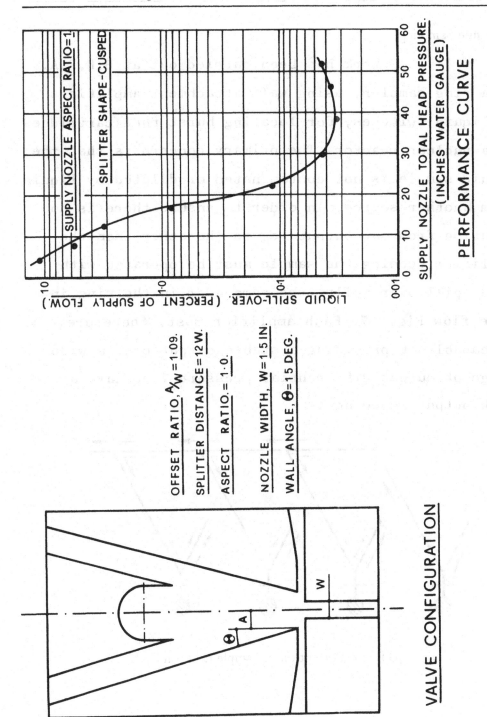

Fig. 16 BHRA Low Head Diverter Valve

c) Liquid Sampling

Work has been carried out at B.H.R.A. on a ring sampler, using wall attachment amplifiers as liquid switches, for locating burst fuel cartridges in a nuclear reactor. The primary concern is that the liquid sample is not contaminated or diluted by liquid from another source. In order to ensure there is no dilution from one output leg to the other then the amplifier feeding the sample must be operated with some spill-over whilst the remainder in the ring induce flow Fig. 17. Each amplifier must, therefore, be capable of providing a stable output over a wide range of output differential pressure i.e. have a high output impedance.

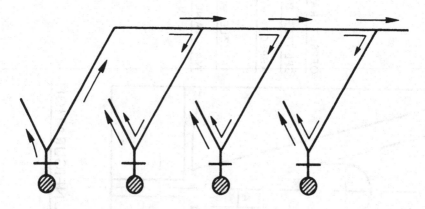

Fig. 17 Prevention of Sample Dilution

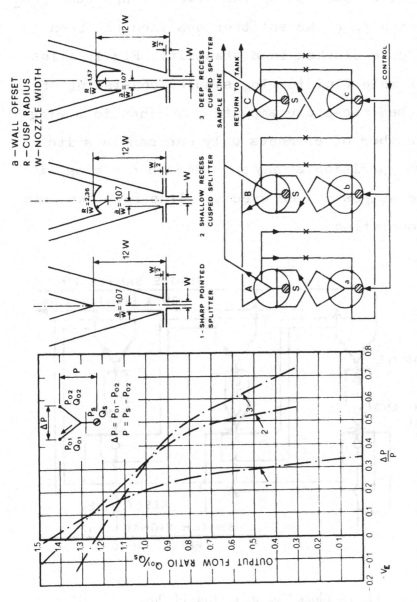

Fig. 18 Unvented Wall Attachment Amplifier - In designing an unvented wall attachment amplifier for a "ring-counter" type sampling system, where dilution or loss of sample is an important feature of the system, each amplifier unit must be capable of providing a stable output flow over a wide range of output differential pressure i.e. have a high output impedance characteristic. The diagrams show the improvement obtained in this characteristic by variation of the splitter geometry. Splitter position is chosen in each case to give blocked load stability.

Fig. 18 (see page 35) shows the improvement in this characteristic obtained by variation of splitter geom etry. In each case the splitter position has been chosen to give blocked load stability. Fig 19 illustrates a simple fluidic ring sampler consisting of three matched two-stage amplifiers. Since in a ring with any number of elements only one sample switch is in the "ON" position at any one time then a ring with only three stages can be used to illustrate the principle of operation.

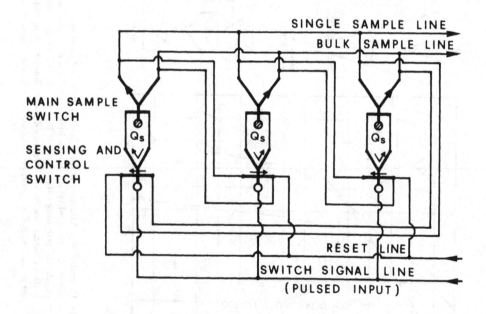

Fig. 19 Block Diagram for Ring of Three

The matched two stage amplifiers con-
sist of control and main switch elements Fig. 20 (see
page 38). For each pulsed switch signal input the
single sample output is indexed in a known sequence,
the bias flow lines providing inhibiting signals to
the gates which are not changing state.

During switching the circulation flow
is assisting the switch signal in the units which
are changing state since the circulation flow must
always be into the reattachment bubble. In order,
however, to ensure complete steering of the switch
signal a bias flow is introduced into the control
amplifier from the outputs of the previous main sample switch. The control amplifier is then being used
to sense the state of the previous sample switch and
during a change of state transmit this information
to its corresponding sample switch. A total bias flow
of around 2 -3 % of the main flow is necessary in
order to prevent switching of the "no change" units.

d) Self Cleaning Filters

Serck Limited [2] have developed a nov-
el self cleaning fluidic filter. This consists of two
filter elements A_1 and A_2 Fig. 21 (see page 39),
housed in the opposite sides of a loop between the

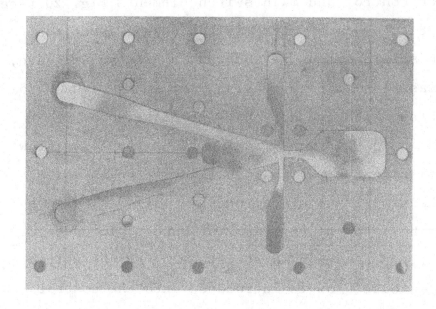

Control Amplifier

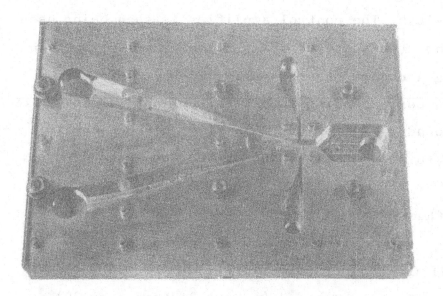

Main Switch Amplifier

Fig. 20

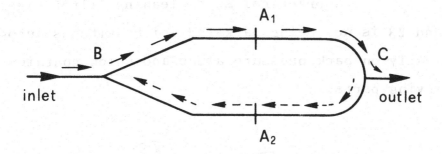

Fig. 21 Self Cleaning Filter System

filter inlet and outlet. At the inlet a wall attachment diverter can direct the main flow stream into
one limb or the other. Downstream of the two filter
elements is a convergence C which creates a back pres
sure and flow in the opposite limb of the loop to
that which carries the main flow. Hence, whilst element A_1 is removing solid particles from the fluid,
the back flow through the other element A_2 is cleaning from its surface the dirt accumulated previously.
When element A_1 has accumulated a sufficient dirt
load the loading on the active output limb of the
diverter causes it to switch making A_2 the active
filter, this having been cleaned by the previous back

flow. A₁ in its turn is now cleaned by reverse flow.

A practical self cleaning filter Figs.
22 and 23 is now being marketed which switches auto-
matically on back pressure alone and which contains
no moving parts.

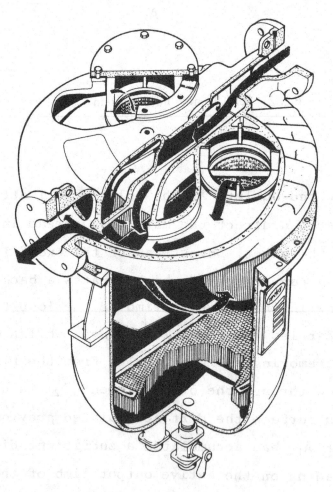

Fig. 22 Serck Self Cleaning Filter

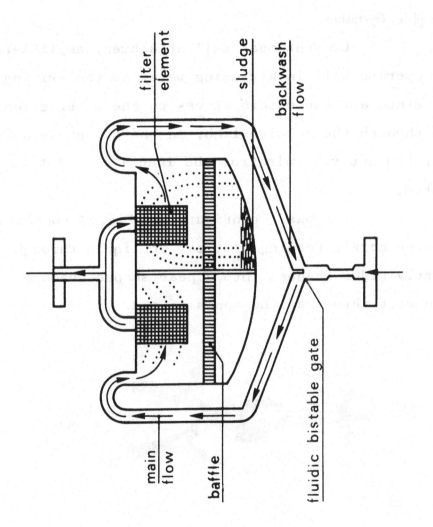

Fig. 23 Serck Self Cleaning Filter

4. TWO PHASE AMPLIFIER

4.1 Principle of Operation

Conventional wall attachment amplifiers do not operate well in air using water as the driving fluid, since air can obtain access to the interaction region through the passive limb. In the two phase amplifier [3] entry of air from the inactive outlet is prevented.

The basic configuration Fig. 24 consists of a power nozzle feeding a stream of liquid through a slightly wider channel into a pear shaped chamber with an exit throat at the opposite end.

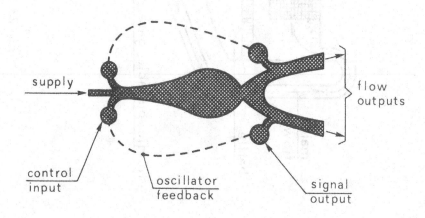

Fig. 24 Two Phase Amplifier as an Oscillator

Two outlet channels lead away from the throat. Control inlets are provided adjacent to the power jet.

The power stream flows along one side of the chamber to the exit throat where it crosses over the active output channel, thereby preventing entry of air through the passive limb. A small part of the stream is peeled off at the throat to provide a circulation in the chamber to hold the power stream onto the chamber wall, giving the amplifier bistability. Switching may be controlled by air, water or selective obstruction of the control channels since in most situations the controls are aspirated well below the outlet pressure.

4.2 Performance

Fig. 25 (see page 44) gives typical pressure as flow characteristics for the amplifier with outlets open to atmospheric conditions. All pressures are sub-atmospheric.

The water signal output is the pressure in the active limb discharging water. As the water flow and velocity increases the pressure falls due to entrainment. The"air signal output" gives the condition in the passive limb i.e. near atmospheric.

Allowing water to be entrained into

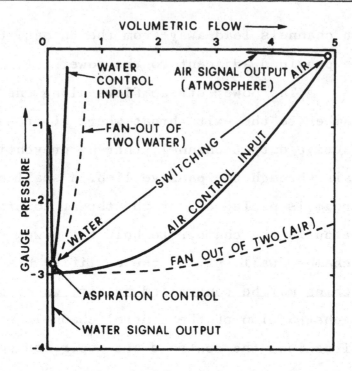

Fig. 25 Two Phase Amplifier - Performance Characteristics

the control input as the flow increase the water control input pressure tends towards atmospheric. Simi-
larly, allowing air entrainment into other control
input the pressure tends towards atmospheric as the
entrained flow is increased.

If identical elements are interconnec-
ted equilibrium conditions will be established at the
cross-over points of their respective characteristics
i.e. water signal output and control input, and air
signal output and control input. Thus, one control
channel is fed with water and the other air resulting

in considerable pressure and flow differentials to
deflect the power stream and switch the amplifier on
reversal of these conditions.

4.3 Industrial Use

The majority of industrial applications
of the two phase amplifier are in the form of oscillators or pulsators providing alternating liquid
pulses from the two outlets. If necessary a number
of elements can be interconnected either in parallel,
series or in combination.

A single element is simply converted
into an oscillator by connecting feed back lines from
each output to the control input on the same side of
the element. The frequency of oscillation is proportional to the flow velocities and inversely proportion
al to the total length of the signal path (feed for-
ward plus feed back).

Fig. 26 (see page 46) shows an indus-
trial liquid agitator with atmospheric discharge. It
consists simply of a two phase oscillator fed with
liquid the output pulses being directed into the liquid filled container, resulting in a high degree of
agitation and mixing.

The device can also be used submerged

Fig. 26 Agitator

as a single phase liquid oscillator for agitation
only, or as a two phase oscillator for aeration as
well. In the latter case each output channel is connected to a stand pipe open to the atmosphere or the
desired gas.

Feedback is taken from the pipe at a
level below the free liquid surface such that an active output channel aspirates the associated stand
pipe level to below the feed back connection. Thus
air or gas is permitted to enter the feedback control
line and switching occurs.

5. VORTEX AMPLIFIER

5.1 Principle of Operation

The vortex amplifier fig. 27 (see page
48) consists of a circular chamber with the outlet
along the axis of symmetry of the chamber. The control flow is introduced tangentially, so that the con
trol stream will impinge on the incoming power stream.
With no control flow the power stream flows radially
from the entry to the centre outlet and experiences
little more than pipe elbow resistance.

When control flow is introduced the
power stream is deflected and forms a spiral vortex.

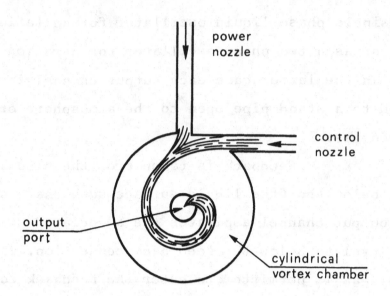

Fig. 27 Fluidic Vortex Valve

The tangential flow velocity at exit is high, due to
the conservation of angular momentum. Centrifugal
forces set up a pressure gradient in a radial direction which opposes and restricts the power input flow.

        The turn down ratio of the valve is
the ratio of the maximum output flow i.e. zero control and a radial power stream, to the minimum output with control flow.

        The vortex amplifier can be used as a
digital two state device or as a proportional amplifier.

        A form of vortex amplifier developed

at B.H.R.A. is shown in Figs. 28 and 29 in which four
radial inlet ports are used for the power flow with
four tangential control ports. Two axial outlets are
used in this particular design, in order to reduce the
resistance to pure radial flow and increase T.D.R.

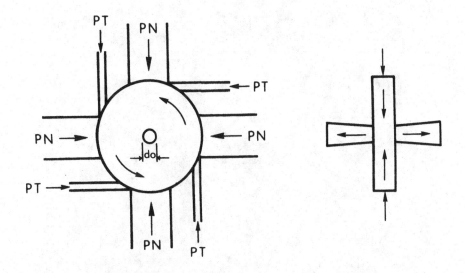

Fig. 28 BHRA Vortex Amplifier

Fig. 29 Photograph of BHRA Vortex Amplifier

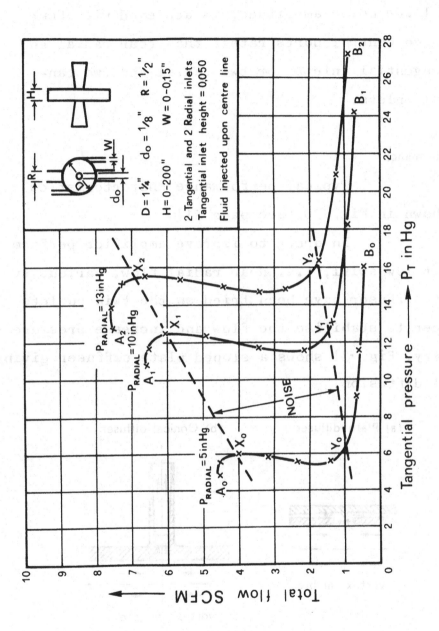

Fig. 30 Vortex Amplifier - Performance Characteristics

Better mixing of the radial and tangential flows and
hence lower noise amplitude, is achieved with four
inlet and control ports rather than four radial and
two tangential inlets, or two radial and two tangential inlets.

5.2 Performance

Typical performance characteristics
are shown in Fig. 30 (see page 51).

In order to improve amplifier perform-
ance at low swirl, i.e. near radial flow, various types of diffuser have been .tried on the twin outlets
in order to stabilise the flow and increase pressure
recovery. Fig. 31 shows a sloped plate diffuser giving
radial diffusion.

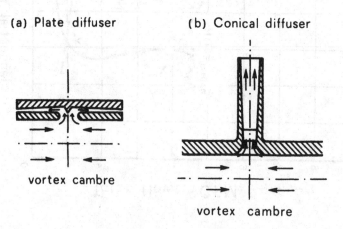

Fig. 31 Vortex Amplifier - Types of Outlet Diffuser

When set to a constant outlet area for radial flow,
equal to the nozzle exit area, the T.D.R. remains
constant at the same value as with no diffuser but
with unwanted noise reduced to insignificant propor-
tions. Increasing diffuser area by moving the plate
axially outwards increases both TDR and noise. This
continues until the radial flow separates and the
flow becomes very unstable. Plate diffusers are very
suitable for amplifiers where low "noise" output signals are required with small overall dimensions.

   The instability or noise associated
with the vortex amplifier arises from the instability of a vortex core issuing into a free atmosphere.
In a region occupying a diameter of approximately one
half of the nozzle exit a strong suction results in
reversed flow and entrainment of the fluid into the
vortex. Fig. 32 (see page 54). An axial rod mounted
symmetrically through the outlet nozzle suppresses
this entrainment and greatly reduces the noise. However, a rod produces large shear stresses in the
swirling flow and also reduces the axial outlet area.
The TDR may be reduced by as much as 40 % but this is
not too important for proportional amplifiers where
stability and low noise is all important.

   Although a conical diffuser produces

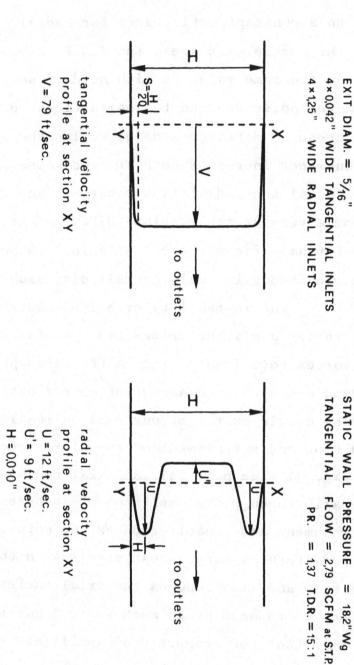

DEVICE SPECIFICATION OUTER DIAM. = 3"

HEIGHT   = 0.208"

EXIT DIAM. = 5/16"

4 × 0.042" WIDE TANGENTIAL INLETS

4 × 1.25" WIDE RADIAL INLETS

$S = \frac{H}{20}$

tangential velocity
profile at section XY

V = 79 ft/sec.

PROBE TAKEN AT 5/8" RADIUS

PRESSURE DROP ACROSS DEVICE = 203"Wg

STATIC WALL PRESSURE   = 18.2"Wg

TANGENTIAL FLOW = 2.79 SCFM at S.T.P.

PR. = 1.37   T.D.R. = 15:1

radial velocity
profile at section XY

U = 12 ft/sec.

U' = 9 ft/sec.

H = 0.010"

Fig. 32 Vortex Amplifier - Exit Flow

much more noise than a plate diffuser it gives a
substantial increase in radial flow and hence TDR
for a given pressure drop when used on a vortex unit
with a low chamber height.

By fitting conical diffusers, the
vortex chamber height can be increased, maintaining
TDR with a very substantial reduction in noise ampli
tude. A parallel throat section of 2 to 3 nozzle
diameters, prior to entry to the conical diffuser,
further reduces noise by stabilising high swirl flows
whilst merely acting as a pipe resistance to radial
flow.

Thus, the use of conical outlet diffusers permits a substantial increase in chamber
height and consequent reduction in noise, making pos
sible proportional amplification with a high TDR.
Vortex units can be divided roughly into two classi-
fications : -

a) Units giving low noise output signals (maximum amplitude 2 % of the inlet pressure)
with TDR up to around 12. These employ either plate
diffuser outlets, or conical diffuser outlets with
a rod coaxially inserted through the outlet centres.

These units are best suited to proportional amplification.

b) Units with TDR up to 20 : 1 with
higher noise output signals (maximum amplitude 10 %
of inlet pressure). These are best suited as high /
low impedance switches but can be employed for propor
tional amplification where noise is of less importance.

5.3 Industrial Application

Less development has been carried out
on the vortex valve than for example the wall attachment amplifier and it is still therefore in a relatively early stage of development. So far as is known,
no details have been published of any industrial application, but development work is still being active
ly pursued in several countries, including the U.K.

6. MOMENTUM INTERACTION AMPLIFIER

6.1 Principle of Operation

The momentum interaction amplifier Fig.
33 (see page 57) is one of the more common types of
proportional fluid amplifier using the vector proper-
ties of fluid streams. A high energy power jet flows
from a nozzle into the interaction region. Low energy
control jets are also directed into the interaction
region, usually at right angles to the power jet. The

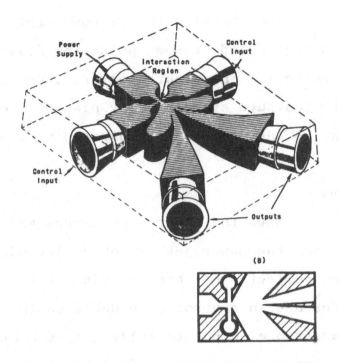

Fig. 33 Momentum Interaction Amplifier

walls are removed between the interaction region
and the outputs, to prevent the flow from attaching
to the sidewalls and making the amplifier bistable.
The momentum flux of the power jet and the forces
exerted on it by the control jets, determine the direction the fluid will assume after it leaves the interaction region. The control jets will apply pressure
forces and / or momentum forces depending on their
position. In general, when the control nozzle is close
to the side of the power jet pressure forces predom-
inate.

The stream diffuses downstream of the
interaction region and at some point the flow is divided and collected in output apertures. Usually there
are only two outputs symmetrically placed, these may
be adjacent to one another or separated by a vent.

6.2 Performance

When the control pressures and flows
are equal then the power jet is not deflected. Each
output then collects the same quantity of flow. A
small difference in control flow deflects the power
jet and causes one output to collect more fluid than
the other. The outputs must be far enough downstream
to take advantage of the deflection of the jet, but
far enough upstream to recover an appreciable portion
of the power jet energy. The gain is described in
terms of the difference of the two outputs and the
difference of the two inputs.

The shape and size of the cut outs in
the side walls has a considerable effect on amplifier
performance. Any fluid not collected by the outputs,
or any disturbances present in the power jet, may be
reflected from the cut out walls back towards the
power jet to produce unwanted feedback. In order to
minimise this, atmospheric vents are fitted to the

cut out region of open type units. In closed, unvented
units the cut outs are interconnected in order to
equalise the pressure across the power jet. The performance of momentum interaction amplifiers is very
dependent on the shape of the jet velocity profile at
the entrance to the outputs. This is changed by vent-
ing the entrainment areas and hence the performance
of closed units differs from that of vented units.

6.3 Industrial Use

Many industrial processes require the
mixing of two or more fluids in a known ratio on a
continuous flow basis typical examples being, the
mixing of oxygen and fuel gas in the correct ratio
to give optimum flame conditions, the blending of
petrols or oils and the treatment of water. The B.O.C.
[4] have developed a novel form of momentum interaction device to perform the function of mixing and
indicating when the mixture reaches the required pro
portions.

The fluidic element Fig. 34 (see page
60) uses the momentum interaction principle. A pair
of jets are arranged at an angle to one another, so
that thay interact in a mixing region. When the rate
of change of momentum of the two jets is equal, the

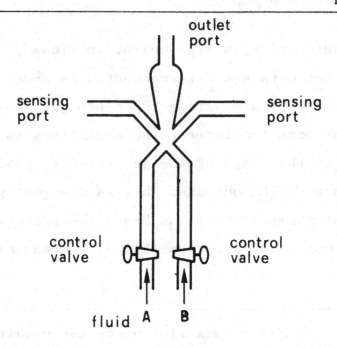

Fig. 34 B.O.C. Fluidic Flow Mixing Valve

resultant jet emerges from the outlet port, which is
connected to the process using the mixed fluids. If
the flow in one of the inlet ports changes, the resultant stream will be deflected towards one of the
sensing ports by the more powerful jet, the deflection being monitored by a differential pressure indicator connected across the two sensing ports. This
provides an indication that the mixture has deviated
from the desired ratio and this may be restored by
operation of the fluid control valves in order to
null the differential pressure indicator.

The device indicates mass flow ratio,
the actual ratio being determined by the relative

sizes of the two main nozzles and the relative densi-
ties of the two fluids flowing through the nozzles,
geometric considerations imposing a limit on the maximum ratio which may be obtained in a single element.
However, if very high ratios are required individual
elements may be arranged in cascade where the two
primary fluids are mixed in the first element, the
resultant mixture passing into a leg of the second
element, where more primary fluid is injected into
the remaining leg to further dilute the mixture Fig.
35. This proceedure may be extended to any number of
stages giving rise to
very high mixture ratios
by a process of succes-
sive dilution. Alterna-
tively, fluids other
than the original pair
may be added along the
chain of elements so
that a number of dif-
fering fluids may be
blended together in a
known ratio on a con-
tinuous flow basis.

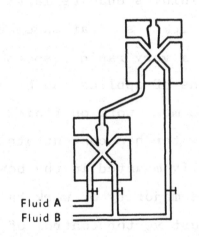

Fluid A
Fluid B

Fig. 35 High Ratio Cascade Mixer

        The device maintains a consistent

mixture ratio over a wide range of flow and may be
designed for gas or liquid operation. Automatic nul-
ling can be achieved by replacing one of the manually
operated fluid control valves by a valve operated
from the differential pressure at the sensing ports,
thereby maintaining a constant mixture ratio over a
wide flow range a feature necessary in automated
process control.

7. THE USE OF FLUIDICS IN MANOEUVRING SYSTEMS

The use of fluidics enables large
flows to be switched very quickly so that manoeuvring
systems can be built having a very rapid response.
Such systems have wide fields of application both in
Aero and Hydrospace. To date most work on fluidic
devices operating with sea water has concentrated on
their use as thrusters usually mounted in the bow of
the vessel. In Aerospace the majority of work has
been concerned with the thrust vector control of rock
ets.

7.1 Hydrospace

The conventional propeller bow thruster
is used for manoeuvring in confined spaces and is generally neither easy to maintain or install in existing
ships. A jet thruster using the reaction force of a
water jet is much simpler. Although the efficiency of
the fluidic version is lower bow thrusters are normal
ly only used for relatively short periods and existing pumping equipment installed for other purposes
e.g. cargo handling, fire fighting ect. can be used.

Bowles, U.S.A., has developed a simple
bow thruster based on the wall attachment amplifier
Fig. 36.

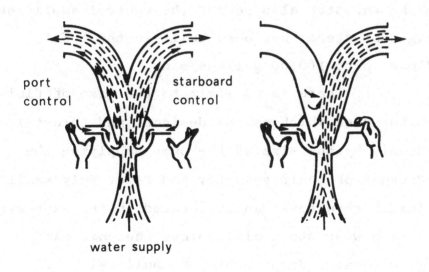

Fig. 36 Fluidic Thruster ('Bang Bang' Control)

When both control ports are uncovered and inducing
air the water stream does not attach to either wall
and divides equally into the two output legs giving
zero nett lateral thrust. Covering one of the ports
causes the stream to attach to the wall and pass out
of the appropriate output channel giving full thrust
in one direction. A disadvantage of this type of
thruster is that since it gives a bang bang control
it could excite bending vibration modes in the ship.

A thruster Fig. 37 (see page 65) using
a proportional instead of a wall attachment amplifier
has also been built by Bowles for the U.S.Navy [5].
A small control amplifier is used to drive the main
proportional amplifier, the thruster. The water supply to the thruster also powers the control amplifier.
Two stage thrusters have been built in the U.S.A.
with flows up to 160,00 gallons/min.

There is no limit to the size of flui-
dic bow thruster which can be designed and thrust is
limited only by the size of the pump supplying the
unit. Because of their geometry and relatively small
size fluidic thrusters can be located in the extrem-
ity of the bow or stern also, since the unit can
usually be located deep within the hull, with its
pump taking bottom suction, the thruster remains

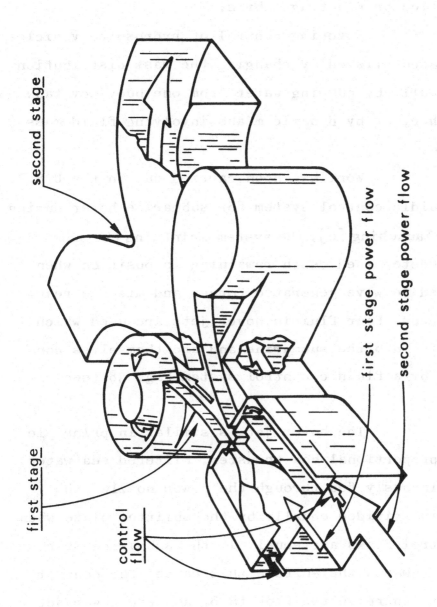

first stage

second stage

control flow

first stage power flow

second stage power flow

Fig. 37 Proportional Fluidic Bow Thruster

operable at very low draught and is unaffected by
surface ice or floating debris.

Hovering control of hydrospace vehicles
can be accomplished by changing the mass distribution
of the hull, by pumping water from one buoyancy tank
to another, or by dynamic means involving fluid mass
ejection.

Work has been carried out in the U.S.A.
on a fluidic control system for submarine hover during
missile launching [6]. The system maintains the submarine
at a predetermined depth remaining in position when
subjected to wave generated forces and missile reaction forces. Four fluidic hover jets are used which
are built into the submarine and are themselves controlled by a fluidic control system Fig. 38 (see
page 67).

The hover jet is similar in principle
to the proportional bow thruster. Filtered sea water
is continuously fed through the power nozzle, the
flow being divided equally by the splitter plate when
the control flows are equal. If the submarine starts
to move towards the surface this causes the control
system to increase the flow to S_1 and the power jet
is deflected upwards a proportional amount, giving
a downward force differential to supply a coun-

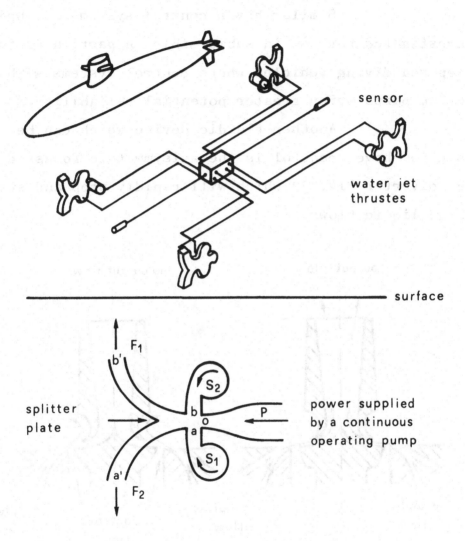

Fig. 38 Fluidic Control System for Submarine Hover

teracting force. A hover jet thruster producing 2400
Ibf has been built and tested at Annopolis U.S.A.

Similar hover control systems are being
investigated for use in submersible in particular for
deep sea diving vehicles where control systems with no
moving parts offer greater potential reliability.

Another fluidic device which can be
used for hover control is the axisymmetric focussed
jet diverter Fig. 39 which will rapidly stop and start
large liquid flows.

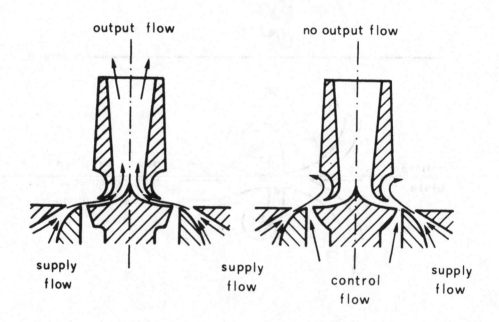

Fig. 39 Axisymmetric Focussed Jet Diverter

In the axisymmetric focussed jet amplifier, first
developed by the Sperry Utah Co., a control flow is
used to switch off or defocus the jet. By preventing
collapse of the power jet the flow passes radially
out through the return channel. Removing the control
signal causes the annular power jet to collapse,
attach to the wall and flow through the output channel. A substantial transitory thrust is obtained
along the amplifier axis at the "on" or "off" switching times, at least one order of magnitude greater
than the steady state thrust caused by the mass flow
from the output of the diverter. This effect could be
of great significance in the control of hydrospace
vehicles.

7.2 Aerospace

A rocket vehicle can be steered by
deflecting the engine thrust, usually by mounting
the engine on gimbals, or by adding auxiliary moving engines. Alternatively, the engine can be kept
stationary and fluid jets used to deflect the rocket
thrust gases to steer the vehicle. The fluid is injected into the thrust nozzle downstream of the engine. The control fluid can be gas bled from the
rocket itself. The valves that control this secon-

dary injection must be capable of withstanding temper-
atures of 5000 ° F or more. In such a severe environment a fluidic diverter should be much more reliable
than a moving part valve. Fig 40 shows a wall attachment diverter manufactured by the British Aircraft
Corporation for such an application using hot rocket
gases. Another interesting application Fig. 41 is the
fluidic jet flap.

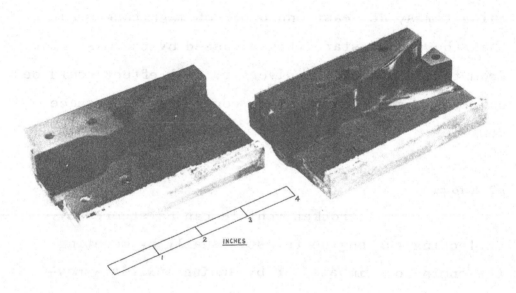

Fig. 40 B.A.C. Wall Attachment Diverter for Use
with Hot Rocket Gases

Air or gas jets can emerge as a sheet from either the upper or lower surface slot fitted near the trailing edge of the wing. The emerging fluid acts as a jet flap, increasing the lift of the aerofoil by as much as 20 times the jet momentum force. The British Aircraft Corporation have developed such a system using a wall attachment amplifier to switch the jet flap. In this application the missile wings are fixed and by switching the jet flap the total con<u>n</u>trol force available is greater than that of the jet flap momentum alone. Alternatively, the jet flap could be fitted to a moving foil and switched fluidically as the foil passes through the zero angle of attack position.

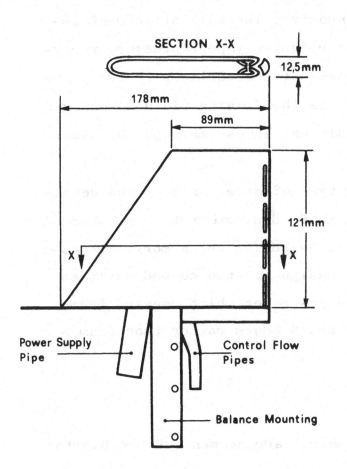

Fig. 41 Fluidic Jet Flap

8. FUTURE DEVELOPMENTS

Large fluidic amplifiers can be used to replace conventional moving part valves in liquid and gas systems in industry. The wall attachment amplifier has however a relatively poor pressure recovery and may not operate satisfactorily with two phase flow e.g. with liquid as the driving fluid discharging to atmosphere, when air has access through the passive output limb.

A new type of diverter has been developed by Bahrton [7], in Sweden, which does not depend on wall attachment for switching and memory. The essential parts of the design are two cusped cavities on either side of the power jet which emerges from a slot formed by two sharp edges rather than from a conventional nozzle.

8.1 Principle of Operation

The general arrangement of the Bahrton valve is shown in Fig. 42 (see page 73).

The use of sharp edges at the exit makes a jet unstable and easier to deflect because of its velocity profile, see Fig. 43 (page 73).

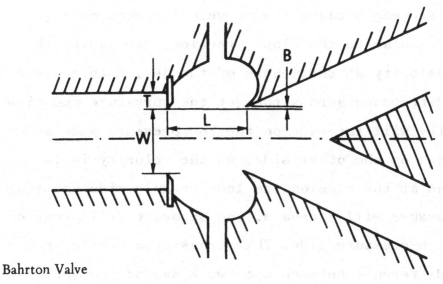

Bahrton Valve

Fig. 42 Characteristic Geometry Showing Edges, Vortex Chambers
and Control Ports. W = Nozzle Width, T = Length of Edge,
B = Setback

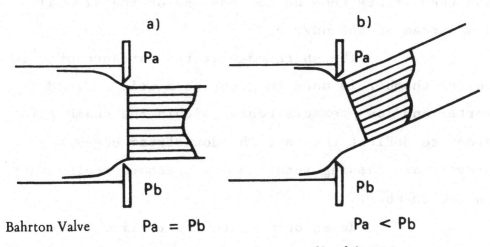

a)

b)

Bahrton Valve   Pa = Pb   Pa < Pb

Fig. 43 Influence of Edges on the Velocity Profile of the Main
Jet: a) Non-Disturbed, b) With a Differential Pressure
Across Main Jet

The sharp edges cause inward velocity components,
perpendicular to the flow direction, decreasing the
flow velocity at the centre of the jet. With a pressure difference across the jet the curvature and flow
velocity will increase on the low pressure side and
decrease on the other side. As the velocity is increased at the edge on the low pressure side the stat
ic pressure will decrease. The opposite will occur on
the high pressure side. This transverse static pressure difference between the two sides of the jet
causes deflection right from the nozzle exit. In the
normal wall attachment amplifier the jet comes
straight out from the exit nozzle with no deflection
and then starts bending towards one of the sidewalls
downstream of the nozzle.

          The sharp edge at the entrance of each
vortex chamber is used to generate a well defined
vortex and low pressure region within the chamber in
order to deflect the jet. The downstream edges also
prevent air passing into the low pressure region with
in the chamber.

          In an open system, with liquid dis-
charging to atmosphere, a small setback is used. With
both control ports open to atmosphere the jet almost
touches the vortex chamber edges and is divided equal

ly between the two output limbs. Closing one control
port, air is sucked out of the vortex chamber and the
jet deflects. Fig. 44.

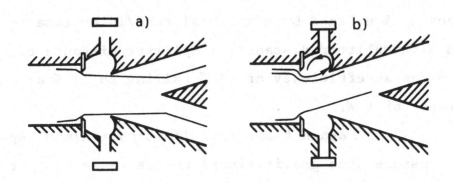

Fig. 44 Working Principle of the Valve a) Control Ports Open,
b) Control Ports Closed, the Upper Closed First.

Part of the jet being peeled off to maintain a well
defined vortex within the chamber. The low pressure
region within the vortex chamber being sealed against
the entry of air through the passive leg. The device
can be switched using either a liquid or gas.

8.2 Performance

           Energy losses in the amplifier are
mainly due to expansion of the jet between the power
jet exit and the entry to output limbs. The highest
energy recovery is obtained with the shortest possible

splitter distance. The minimum splitter distance obtainable, before spill over occured into the passive leg, was 2.5. W.

Fig. 45 shows the variation of efficiency η % defined by exit total head/inlet total head with splitter distance. A splitter distance of 3 W gives an efficiency of 95 % falling to 65 % at a distance of 6 W.

Again there is a limiting minimum supply pressure when gravitational forces cause the jet to spread and spill over into the passive leg.

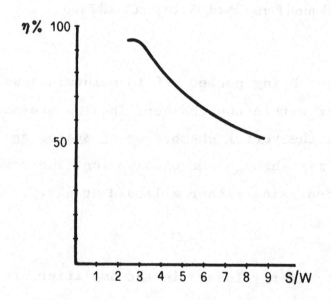

Fig. 45 Pressure Recovery as Function of S/W

9. CONCLUSIONS

Large fluidic elements offer attractive possibilites in a wide variety of industries, in particular for flow control in the chemical and nuclear industries [8] and for manoeuvring systems in Aerospace and Hydrospace [9] .

Non moving part devices should be cheaper and more reliable particularly in a poor environment or hazardous area. Fluidics is inherently safe, does not call for any specialised maintenance, and offers several advantages over moving part valves. The moving seals of conventional valves are eliminated and valve actuators are not required. In particular, rapid switching of large flows by fluidic diverters cannot produce large water hammer pressures, a problem with conventional valves. Their main disadvantage is that of power consumption, some 50 % higher than for most conventional moving part valves.

Although hardware costs of fluidic equipment are low, development costs are high. Both in the U.K. and U.S.A., the majority of financial support has come from government contracts. Successful applications are likely therefore to come as a result of

this work but this may take another two or three
years.

REFERENCES

[1] Bain D.C. and Baker P.: "Technical and Market Survey of Fluidics in the U.K." pp.135, 1969, British Hydromechanics Research Association, Cranfield.

[2] Montgomerie G.A. and Floyd T.J.: "Heavy Current Fluidic Devices" Paper K4, Second Cranfield Fluidics Conference, 1967.

[3] Bauer P.: "A two phase fluidic element and its application" Fluidics Feedback Vol.4 no. 3, pp. 68-71, March 1970.

[4] Miller R.E.: "Fluidic Gas Mixing" Paper T7, 3rd Cranfield Fluidics Conference 1968.

[5] Dexter E.M.: "Applications to solutions of control problems" Fluidics Quarterly, vol. 1 no.2, pp. 103-110, January 1968.

[6] Buck J.R.: "Pure fluidic control systems for submarine hover during missile launching" Naval Engineers Journal, pp. 759-766, October 1968.

[7] Bahrton S.: "A new type of fluidic diverting valve" Paper A4, Fourth Cranfield Fluidics Conference 1970.

[8] Baker P.J. and Bain D.C.: "Large scale fluidics in process control" Chemical Processing 15.10. Process Instrumentation and Analysis Supplement October 1969.

[9]   Bain D.C. and Baker P.J.: "Fluidics in Marine
              Engineering Part 3 – Heavy Current De-
              vices" Marine Engineer and Naval Archi-
              tect 92. 1125, pp. 513-519, December
              1969.

CONTENTS

Printed in the United States
By Bookmasters